This Notebook Belongs To

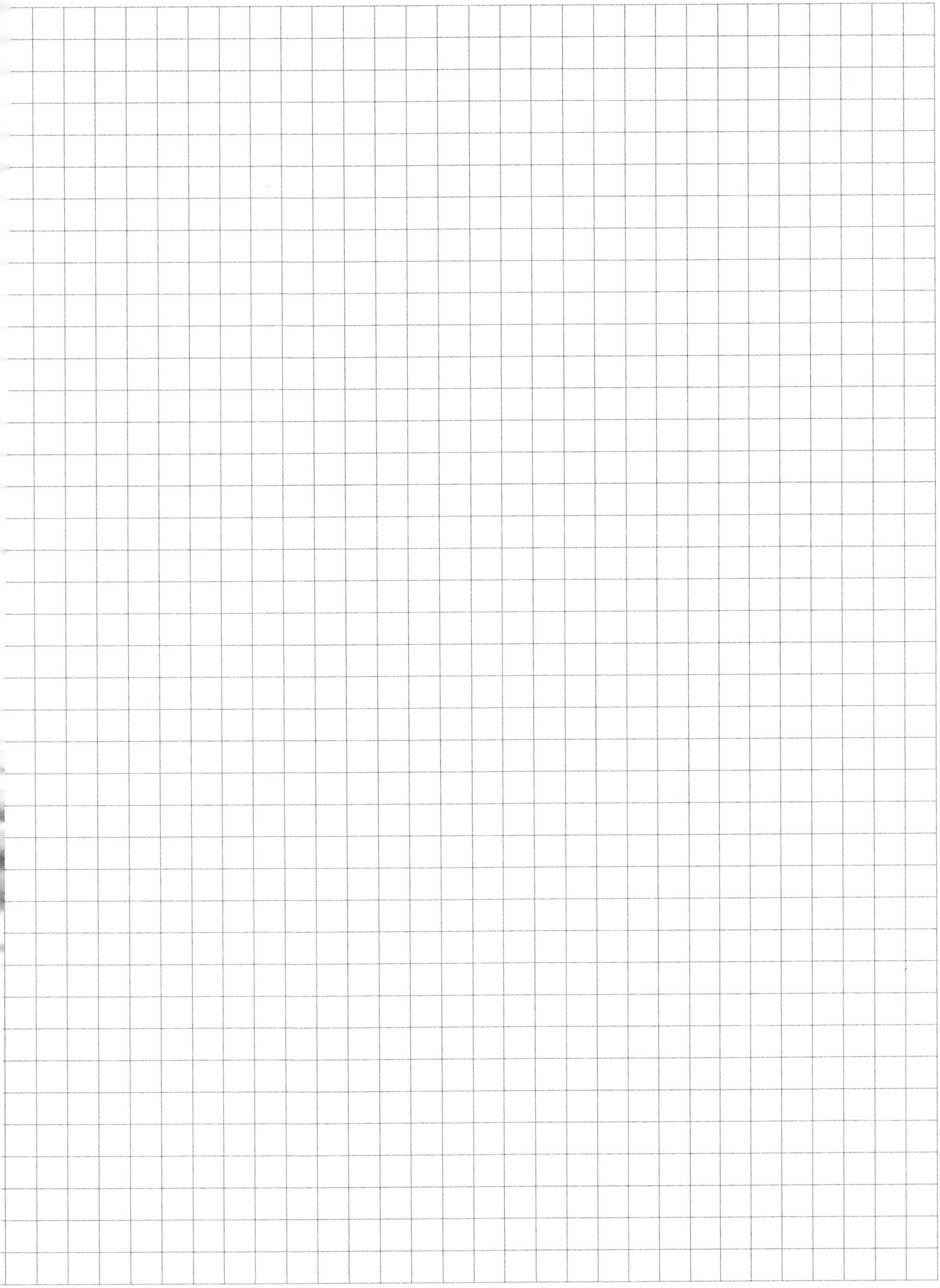

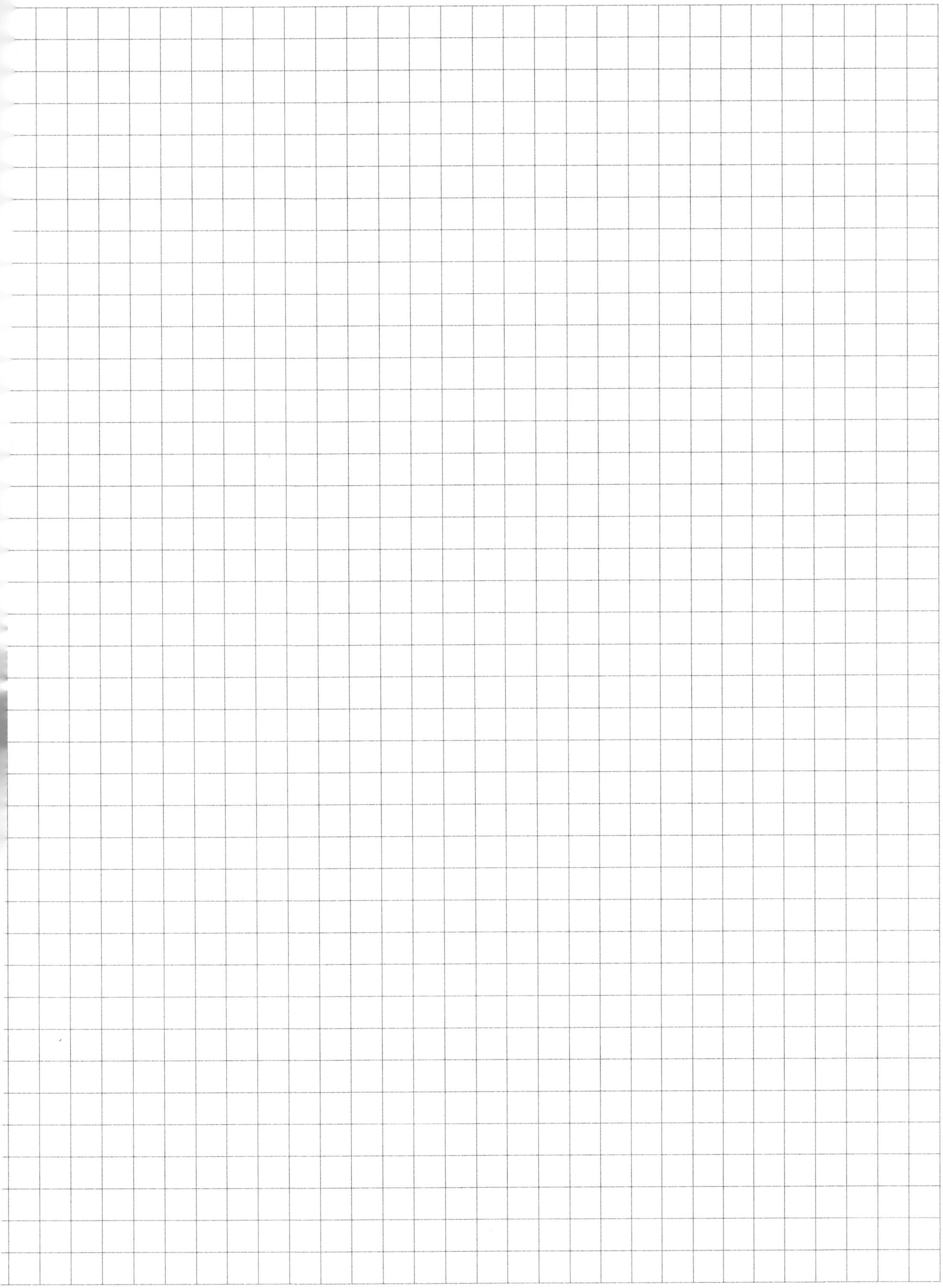

Made in the USA
Coppell, TX
04 November 2026

11287996R00057